Table of Contents

INTRODUCTION

"The next World War will be over water"
> – Ismail Serageldin

While Ismail Serageldin's quote may seem unbelievable to some, others may consider this a problem that only exists outside of the United States. Despite the doomsday feel to that statement, the simple truth is this - the world is running out of easily accessible fresh water. Highly developed countries face this problem as much as developing ones as location and geography are the biggest factors in water resources.

While the United States is very blessed with tremendous water resources, we are not immune to the problem. According to the U.S. Government, 36 U.S. states are already facing water shortages or will be facing water shortages within the next few years. The ten biggest cities currently running out of water:

1. Los Angeles, CA (Population 3,700,000)
2. Houston, TX (Population 2,100,000)
3. Phoenix, AZ (Population 1,445,000)
4. San Antonio, TX (Population 1,327,000)
5. San Francisco, CA (Population 805,000)
6. Fort Worth, TX (Population 745,000)
7. Las Vegas, NV (Population 583,000)
8. Tucson, AZ (Population 520,000)
9. Atlanta, GA (Population 420,000
10. Orlando, FL (Population 240,000)

America has a problem. It is a complicated problem with a challenging solution. Americans have taken access to healthy drinking water for granted. Potable water is a limited resource. Our demand for it cannot be unlimited. Providing drinking water has unseen costs that consumers often don't realize. Most people don't think about how we come across water. It is cheap, clean and infinite at the end of their faucet. The truth of the matter is, providing potable water takes energy to process and transfer, consumes resources and requires infrastructure. Treatment centers have operational costs for staff, equipment and land, pumps require a tremendous amount of energy to distribute the water over a citywide network or from water sources and pipes are constantly in need of repair and replacement.

The reason that people don't comprehend water's true cost is that municipalities heavily subsidize this cost. Having access to clean water is a vital part of our health, business and our civilization. Clean water goes beyond providing us with the sustenance. It prevents infection and disease spread. It controls sewage and waste disposal. It allows us to clean ourselves, water our crops, manufacture our goods, cool our buildings and of course, be consumed for sustenance.

We have a problem and the time to act is now: it is estimated that California and New Mexico only have a 20 year and 10 year supply of fresh water left respectively. Eight states in the Great Lakes region have signed a pact banning the export of water to outsiders - even to other U.S. states.

The Cause

While we know based on current usage and resources that there is going to be a shortage of water, determining the cause is a lot murkier. Population increase, per capita usage increase of water, infrastructure degradation, weather patterns, increased power demand and agriculture are all contributing factors.

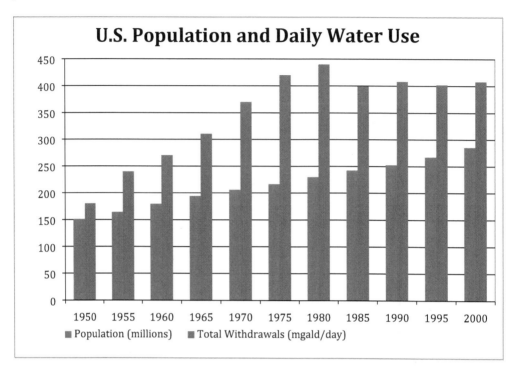

Looking at the above graph, you can see the good news. There has been a steady population increase in the US since World War II, but since 1980 water withdrawals had balanced out. The beginning of the environmental movement the creation of higher efficiency toilets and other fixtures caused much of this reduction. In 1992, congress took this a step further by passing the first Energy Policy Act[2]. It required greater efficiency in power plants, the development of renewable energy sources and limits on flow and flush rates of plumbing fixtures. EPACT 1992, as it is commonly referred, required that toilets not exceed 1.6 gallons per flush, and flow rates of showerheads and faucets to not exceed more than 2.5 GPM at 80 psi.

Looking deeper into that chart you will also see that the water withdrawals are slowly increasing again.

---

[1] Source US Geological Society *Estimated Use of Water in the United States 2000*
[2] H.R. 776, cited as the Energy Policy Act of 1992, 102nd Congress law 486

According to a US Geologic Society report, roughly 64% of all water withdrawals in the U.S. is used for the purposes of power generation.

What is the main driver of water use? Unfortunately, there is no simple answer. Population, power demand, manufacturing and agriculture all play a part. In the United States, the average water use per person, per day (Water Footprint) is about 70-100 gallons and our population (and water use) is rapidly increasing. According to the EPA, the United States population increased nearly 90% from 1950 to 2000. During that same time period, water demand increased 209%. Global population plays a role in this as well due to United States exportation of goods requiring water in their production (Virtual Water). Global population hit 7 Billion in 2012. It was 6 Billion in 1999. By 2030, the world population is projected to be 8.3 Billion.

Drought also plays a large role. No matter which side of the fence you're on regarding Climate Change and Global Warming, the fact remains the U.S. interior west is now the driest that it has been in 500 years according to the U.S. National Academy of Sciences.

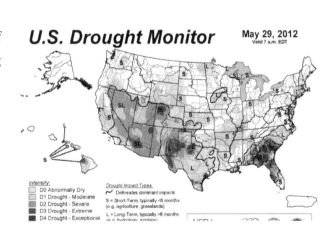

Exacerbating our water problem is the current state of our aging infrastructure. The EPA estimates nearly 50% the United States piping will be classified as either poor, very poor, or elapsed by 2020. In Oklahoma City, 685 breaks occurred between July and September of 2012. According to the EPA, there are approximately 700 water main breaks per day and the longer we wait to replace these existing systems, the higher the cost. The EPA has estimated the cost to upgrade the United States water infrastructure over the next 20 years between $750 million to $1 trillion. Without Federal funding, these costs will be passed along to the consumer in higher water and sewer rates. For example, in October of 2011, Chicago Mayor Rahm Emanuel launched a $4.1 billion initiative to rehabilitate their crumbling water utility infrastructure (it is estimated Chicago has over a thousand miles of water line at least 100 years old). As a result, it is expected by 2015, the people of Chicago will pay twice as much for water.

To exacerbate the problem, water infrastructure cannot be replaced immediately. Streets need to be closed, traffic needs to be diverted and trenched need to be dug before you even starts replacing the pipes. Mot cities estimate that it would take approximately 20 years to replace their water infrastructure because they cannot close more than 5% of their streets at a time.

Fresh water is limited. There exists the same amount of water on Earth today as there was billions of years ago. While it covers most of the planet, only 3% is freshwater (mostly ice), less than 1% is readily accessible for human use and less than 0.007% is available to drink. We can "make" new water through technology (desalination, rainwater harvesting, reclaimed water, and reused water) but in reality there exists no better payback than conservation. Additionally, technological means and methods of creating new water can take decades from planning, funding, design, construction to use while conservation is nearly immediate by comparison. Moreover, many cities, counties and utilities offer conservation incentives resulting in even quicker paybacks.

Some scholars and economists have tagged water as next energy crisis. If this is true, we are not ready for it. Do an internet search for "Water Auditors" and you will instantly understand. There exists no list of Building Water Auditors from which to choose and definitely no references to a workshop offering to teach this skill. This workbook will change both.

A Building Water Auditor not only contains the skills to walk a facility and quickly identify water saving opportunities for the owner, but also to create rewarding work with your best clients, and build relationships with new ones. All the while, this is accomplished while saving them operational expenses, conserving our most precious natural resource and creating profitable work for your business. Most building owners are currently unaware of their potential savings possible, rebates available and how quickly this unobtrusive work can be implemented from audit to installation with minimal investment. A Building Water Auditor can successfully lead them through this process.

With utility rates skyrocketing, public awareness of water scarcity and sustainability, and cost-effective, reliable, water efficient products on the market for all fixtures and equipment, it is easy to see why virtually every existing facility in existence (estimated at over 70 billion square feet in the United States and growing) is a new opportunity.

CHAPTER ONE – WATER AUDITS (WHO WHAT, WHERE, WHEN, AND WHY)

"When the well's dry, we know the worth of water"
– Benjamin Franklin

Water Auditing is a broad term that can carry different definitions depending upon with whom you are talking. Before we dive into to the nuts and bolts of performing a water audit, we must first be sure we are in agreement with what a water audit is and what type of water audit this book offers those who would like to learn.

What is a water audit?

A water audit is an on-site status report of a building's water infrastructure. Sometimes, this includes the surrounding property of the building as well. It includes a detailed review of the water consumption devices or system in a building such as plumbing fixtures, equipment, and systems.

Basic Examples of Water Consuming Devices

Plumbing Fixtures	Equipment	Systems
Water Closets	Cooling towers	Cooling
Faucets	Ice makers	Rainwater
Showers	Dishwashers	Irrigation

After the on-site review, a detailed water audit report is delivered which contains:
- Equipment and systems reviewed
- The status of the fixtures
- Suggestions of water conservation opportunities
- Potential rebates or incentives available.
- Payback Information

The payback information in the water audit report is <u>crucial</u> in an effort to allow the owner to decide on whether or not to fund a particular suggestion. While conservation may be a goal for the owner, there will always be monetary limits to what they can do.

There are indirect cost savings that need to be taken into account in the auditing process as well. Though they might seem secondary at first glance, they must be factored in to gain the whole picture. While we are focused on conserving water, the fact remains that in saving water, we also save wastewater. Additionally, if we are saving hot water, we are also saving hot water generating fuel. These secondary factors add up quick as in most places, the cost of wastewater is greater than or equal to the cost of potable water.

Another factor to consider is maintenance. Newer products tend to require less routine maintenance. This is a lot harder to quantify in cost analysis, but should be taken into consideration when speaking to the owner. For owners, there is also a cost to of not doing anything as well. By not implementing some of the suggestions, the owner will be subject to ever increasing utility costs at their current use rates.

Using the payback worksheets contained in this workbook will allow for you to capture all of these real costs in your payback calculation and, as we all are aware, the shorter the payback period, the more likely the owner will choose to proceed with the suggested upgrades.

Who performs water audits?

Three separate groups – water conservation companies, utilities and plumbing contractors, generally perform water audits.

Water Conservation Companies

Water conservation companies are typically a third party professional reference for a building owner or user and charge for their work. It is not uncommon for this work to cost an owner up to $0.01 a square foot. Usually, the water audit is very thorough and not only provides feedback on what can be improved water-efficiency wise within the building, but also outside irrigation and the status of all other systems that do not offer any opportunity for conserving water.

Water conservation companies typically don't self perform their own work. Should the building owner or user decide to move forward with the upgrades presented, the water conservation company will either bid this work out publicly as another service (and fee) or the owner or user can elect to bid or perform the work themselves. Unfortunately for the owner, a publicly bid project will require documents to be produced (another fee) and they may end up with a contractor they have never done business with before as the low bidder. Also, the products recommended by the water conservation company may or may not be exactly what the owner desires and the low bidder may elect to substitute their own least expensive option for their proposal.

As a result, the effectiveness of the final product may not be as intended. In fact, functionality and additional maintenance is often placed second behind pricing and installation efficiency in the public bid arena and neither of these is desirable to an owner.

Utility Companies

The second group is private entities such as water and wastewater treatment facilities. This can also include counties, cities and states. Water audits performed by these groups are typically provided as a free service to the community, but are not as thorough as those offered by water conservation companies. The reason for this is they are focused on reducing the amount of infrastructure required for water and wastewater and not generating a profit. As a result they focus on only those items that offer water saving opportunities and have no expertise in fixtures, equipment or systems.

Similar to the water conservation companies, these entities do not perform their own work and will leave it up to the owner to pursue the water conservation opportunities presented. They do not offer any project management capabilities and suggest no fixture specification beyond their flow and flush rates. For an owner to perform this work, they must either perform the work themselves or hire a general contractor to prepare documents for public bid. The concerns mentioned above regarding publicly bid projects remain the same in this scenario.

Plumbing Contractors

The third group is professional plumbing contactors. Once a plumbing contractor has successfully learned how to audit a building and prepare the report, they can self perform the work and remove the middleman. Plumbing contractors are adept at computing accurate installation costs. Plumbers have more expertise in choosing products and materials proven to be reliable, functional and low–maintenance. Once trained in water auditing, they are experts in quickly identifying only those opportunities for an owner to consider that offer payback periods within a set of payback parameters. For instance, if an owner has a requirement that all upgrades must offer, at a minimum, a 3-year payback, the contractor will focus only on those items, in their experience, during their water audit. This saves the owner time as well as the contractor.

When should a water audit be performed?

Now. Just about every building offers potential water savings. The time is now for owners and contractors alike to perform building water audits. A quick internet search will let you know if incentives plans are currently offered in your area and for the most part, it's first come, first serve until funding for the year is allocated.

Due to the current state of our nation's water and waste water infrastructure, utilities can no longer put off essential repairs, regardless of incentives. In other areas, they are also spending funds on expanding their systems to accommodate increasing population and upgrading to fulfill state and federal mandates for waste water treatment.

Some of the more water-poor areas are adding more infrastructure in the form of reclaimed water and non-potable water systems. Inevitably, these costs are passed along to consumers in higher rates for their service. Today, we're seeing businesses absorbing as much as 30% increases in their sewer bill from one year to the next with the water and hot water fuel expenses not far behind. Utility costs will always depend strictly upon your location in the nation and its current situation regarding weather, infrastructure status, local regulations and utility pricing. To be blunt, most owners would be wise to invest in product upgrades and enjoy their new equipment than stick with what they have and use this same money to pay for increased utility cost.

Like utility costs, labor, commodity and equipment pricing are continually rising as well. It is fair to say the longer an owner waits to make water efficient improvements, the more expensive it will become.

Where should a water audit be performed?

Everywhere. Almost <u>every</u> existing building has potential water savings, including new construction. Unless the building was built to LEED Platinum guidelines or is a netzero structure, there exists the potential to increase its water efficiency. While there exists more water saving opportunities in older structures compared to newer ones, the cost of installation in older buildings is typically more when compared to newer versions. Additional considerations should also be made to buildings that have more operational days like hospitals and public buildings. With more operational days, there are faster payback periods.

> ### *LEED Platinum and Net-Zero Buildings:*
> LEED and Net-Zero (NZB) are terms used to identify green buildings. Although there are incentives for each to include water reductions, there can still be opportunities. NZB's focus almost exclusively on energy conservation and although there are minimum requirements for LEED now, older versions required no water conservation features.

Point being, all buildings in existence have potential but each will require an audit to determine the all important payback period.

Why should a water audit be performed?

Perhaps a better question would be why not? There are obvious environmental advantages as well as financial incentives to having an audit performed. By retrofitting a building with new, efficient equipment, an owner reaps two main benefits. They get new equipment and they reduce their overhead. Positioning your company to spend less on overhead for years to come is a competitive advantage.

It is, in fact, good for our environment and gives us a chance to do something for generations to follow. It is also good for the world. Our water isn't just used directly and only by us. Virtual water—our virtual water—is used in manufacturing and growing everything we produce and export in the United States. In particular, a lot of our goods go to areas of the world where they never had or have lost the luxury of water to produce these same essential goods economically. Preserving our water resources ensures we can continue, and perhaps even expand, this role in the world.

> ***Virtual Water Defined:***
> Water used in the production of a good or service. For example, 1,300 liters of water are used at the farm to produce 1 kg of wheat, that can be consumed anywhere. Therefore, when 1 kilogram of wheat is exported, 1,300 virtual liters of water resources are also exported.

Water audits are also obviously good for our customer. During a time of unending utility rate increases, we offer a solution that not only saves the owner from absorbing these additional operational expenses immediately, but also leaves them with new, more efficient and reliable fixtures and equipment.

For the plumbing contractor, water auditing gives us the ability to differentiate ourselves from others and show that we are service provider and community environmental steward. Best of all, it adds value to your company as this is a service you can offer every building owner in your area and create new friends, contacts and profitable work.

CHAPTER TWO – HOW TO PERFORM A WATER AUDIT

"Water is the only drink for a wise man"
  – Henry David Thoreau

Performing a water audit is a step-by-step, systematic process. Once the process is known, it can be applied to any building, anywhere.

The first step is collecting data. Before stepping inside a building, you should be armed with as much information as possible. This includes:
- Current utility rates
    - Water
    - Sewer
    - Electric
    - Natural gas
- Historical pricing (if available)
- Incentives
- Operations Information
    - Number of days in operation
    - Number of Employees
    - Number of visitors
- Blueprints

By collecting this information on the front end, it allows you to look for specific fixtures or equipment that are easy payback targets.

Most facilities track all of their utility rates closely and they can either print off this data for your use or email it to you for analysis. As a water auditor, you will need to know all of this data in order to properly calculate paybacks because the use of each of these may be reduced when saving water.

In most areas, sewer rates are determined by water usage. Therefore, every gallon of water saved, results in a gallon of sewer fees saved. Even better, saving sewer fees in most all areas saves the user more money than saving water. Due to the energy intensive nature of sewage conveyance and treatment, sewer fees are typically two to three times more expensive than water.

Many people also overlook another very important part of the equation - energy use. When the amount of hot water is reduced in any application, the amount of energy or gas used to heat that water is reduced. Most buildings heat their hot water either through direct electric or natural gas hot water heaters, or indirectly through steam to hot water converters.

Collecting historical data is essential for determining utility cost trends and establishing anticipated utility annual increases when calculating water efficiency paybacks. While past increases do not guarantee the accuracy of future increase predictions, it does give you the best idea that is not available by any other means. Contacting the utility companies directly may help in forecasting future increases as many are required to announce future rate changes. However these rates are obviously subject to change and can be impacted greatly by unanticipated repairs, upgrades, codes, guidelines or drastic changes in area population or weather.

With so many variables, determining the correct percentage of annual utility increases is difficult, however one thing is for sure: it is not zero. These rates will undoubtedly continue to rise annually; it just needs to be identified as accurately as possible and communicated to the owner. The best way to do this is to come to a consensus with the owner and yourself (and the utility if a rebate is part of the proposed work). Without a consensus, using the calculator should allow you to change the utility rate forecast easily. This will allow you to bring in multiple calculations based on no change, low-end estimate and high-end estimates.

The second step is to collect specific building data. In addition to their water and sewer bills, it is equally important to find out the facility metering set-up. Do they have more than one meter servicing the facility, do they have a separate meter for irrigation, does the facility have any sub-meters? Each of these items will assist you in identifying auditing focus areas and for tracking post-installation effectiveness. Additionally, obtaining the latest building blueprints is essential, especially for larger facilities. You will need to use these blueprints as a roadmap for the audit (ensuring nothing is missed) and an installation guide should the work proceed.

Other key data includes occupancy and building use. These figures directly impact usage and payback calculations.

Important questions to ask:
- How many days out of the year is the building in operation?
- Is it used during the summer the same as in the winter?
- How about evenings and weekends?
- Are there separate users during those times?
- What are the buildings full time occupants and part time occupants?

Using a hospital as an example, a staff bathroom on a patient room floor is used 24 hours a day, 7 days a week by all staff while a public bathroom only available (and used) during visiting hours have vastly differing use rates. In short, fixtures and equipment used more often provide faster payback periods so it's important to accurately identify all users of the facility and their occupancy durations. In the end, an average is

According to LEED: The average male commercial building user uses the urinal twice and the toilet once during a standard eight-hour day.*

* Actual results may vary

determined by water auditor based upon this information along with other helpful guides such as LEED and IAPMO usage guides.

Once this information is collected, you are ready to perform the audit. It is imperative to be diligent, thorough and organized throughout the process. Using the blueprint as a guide, make sure every room on every floor is reviewed.

A common mistake during the actual audit is to assume rooms are alike and count them the same. Even if the actual facility personnel advise you they are the same, look at them all anyway as even they can forget or be surprised of upgrades or fixes performed without their knowledge or that have been forgotten. Sometimes the smallest difference can make a big impact on the work performed. If there is atypical piping in a couple rooms or access to the water is different, it might take more installation time.

The importance of seeing everything is because your room-by-room audit will be used as not only an installation guide, but also a purchasing guide. For plumbing fixtures, the spreadsheet included in this workbook has stood the test of time in being a reliable guide for this purpose. It is also in an excel format on the disk and can be customized to match your particular auditing style. **Accuracy is critical,** as miscounts will cost time and money correcting them during construction.

In addition to the blueprints and fixture spreadsheet, you will need to have a set of tools to accommodate any situation you many encounter. Make sure all toolbox items are in good working order and you have extra sets of batteries and forms. The last thing you want to do is go back to the office to look for a working flashlight.

> **The Audit Toolbox:**
> - Question List
> - Rebate Forms
> - Measuring Bucket or Bag
> - Stopwatch
> - Flashlight
> - Camera

Regardless of whether or not you are charging for the water audit itself, time is money for you and your customer. It makes no sense to review systems that the owner has no interest in improving or will not offer a reasonable payback period. For example, you may have outstanding historical documented savings on new dishwashers, however the owner may have no interest or funding available for making this investment. Unless you have some unbeatable deal, skip it and move on.

Another example is that the owner may be interested in utilizing rainwater harvesting for all of their maintenance building urinals, but from your experience you know that there are no incentives and payback periods that are typically decades long. In these cases, it might be best to let the owner know that the payback period is extremely long, there are code issues, and there are concerns from waterborne pathogens.

The last major concern before you do your audit is to make sure you can access all areas prior to the water audit. Some buildings may require additional coordination regarding security clearance or touring certain areas before or after employee work hours.

Onsite

The main objective during the audit is to effectively use your time while not missing anything important. You do not want to miss anything that you will need to perform the payback calculations or potential future installation. One of the hardest things to do in big buildings is to stay focused. Often times you are ticking off the same boxes ten rooms in a row and not paying attention to details.

While auditing onsite, remember that you need to:
- Take notes. Anything different needs to be noted.
- Ask questions. The staff can always give you better operational details of what is working, what isn't and what they don't like about the plumbing.
- Take pictures. Document existing conditions as best you can. It also helps you make the sale when you can show examples of the equipment in question when the building manager or owner might not be familiar with the item in question.
- Be wary of changes not reflected on the drawings you are given. Undocumented remodels or upgrades are common.
- Test flow rates. Don't just trust the label. Flush rates are a little more exact, but there could be a number of factors impacting the flow of water in faucets and showers.
- Test the equipment. Make sure the equipment is in good working condition. This will help you identify items that might need to be replaced for maintenance sake.

A return trip should not be necessary upon completion. If possible, try to determine who the final decision maker on approving this work is. Once this is known, find out what this person is mostly interested in, their needs and their importance: paybacks, green training, new products, reliability, etc. and match your efforts to their desires in an effort to make the sale.

CHAPTER THREE –CALCULATING PAYBACK PERIODS

"Water flows uphill towards money"
  – Anonymous

Effectively calculating the payback periods are the key to making a sale. While you may experience many owners and facility managers interested in sustainability, it needs to be understood that being sustainable to most means having their same job, this same time, next year. In order to maintain their employment, they need to pay attention to the bottom line and that means paying attention to the payback period.

The term, "payback period" may also mean different things to different people. Care should be taken in making sure you and your customer are working towards the same definition. For our purposes in this workbook, the payback period is the amount of time it takes to pay for the improvement installation through future savings. In simple terms, if an improvement costs $100 to install and the forecasted savings is $20 / year, the payback period is 5 years.

Costs

To determine the costs, you need to know all of the variables that are included:
- Labor
- Materials
- Equipment,
- Mark-ups

Calculating your labor rate should be a fairly easy function. Upon returning from the audit, meet with your estimator to coordinate labor estimates for each fixture type and piece of equipment.

Be sure to separate out cost forecasts to each type of installation. You will need to calculate payback periods for each fixture and equipment separately. This will allow for greater flexibility when taking your audit information to the owner. You might decide to combine quick payback features with longer payback features in order to bring the average in at a reasonable level.

Do not forget to incorporate additional materials and mark-ups individually as well.

> **Example**:
> An owner has a three-year payback threshold on all work performed. If your calculations estimate an 18-month payback on toilets, but a four-year payback on faucets and sinks, it might be in your best interest to combine all estimates into the entire bathroom. The total bathroom payback period would average out to 2 ½ years and fit within the owner's paramaters.

Savings

Once you know the costs, you need to calculate the savings, which include:
- Gallons of water saved
- Gallons of wastewater reduced from sewer charges
- Applicable rebates
- Electrical usage reduced from water conservation or
- Natural gas usage reduced
- Avoided maintenance
- Utility cost increase savings.

Determining the savings by fixture will help identify each individual payback period. Calculating the *actual* savings will always be difficult and inexact because much of it depends on usage, forecasting utility rates, water pressure and so many unforeseeable variables. Many assumptions are going to have to be made for the forecast to be reasonable. The payback calculator does the majority of the work here.

In the Payback Period Calculation Spreadsheet, you will enter in all of the associated data for each fixture. To determine the gallons of water saved, you need to first enter in data for existing fixtures, then the new fixtures. Make sure that you have all of the necessary information from the audit.

Natural gas or electricity usage will be determined by calculations on amount of hot water used. The spreadsheet has a default assumption of 78.6% hot water used on faucets and showerheads. Once the utility rates are entered in for electricity and natural gas, the spreadsheet will determine the amount of each resource saved automatically.

Usage Rates

Use rates based upon audited occupancy and building use are also considered. As professional contractors, estimating the upgrade costs is a routine event performed continually and highly accurate. The savings is not as exact simply because it is based upon numerous variables. The utility rate increases are an estimated forecast based upon past trends and future anticipations. Use rates are determined from occupancy data, building operational hours, LEED guidelines, IAPMO recommendations and common sense.

Before using the payback calculation form, a quick review of its creation and how it works is necessary. Most utility companies offering rebates want to know the cost of the work and potential savings. Almost none of the utility companies give you a form to use to make this calculation. Others give you a form, but do not include the flexibility to calculate real savings for rebates, maintenance avoidance and hot water creation fuel. Still others are only applicable for plumbing fixtures and not entire systems or become cumbersome when calculating multi-use fixtures such as dual-flushometers. LEED has a nice water savings calculator but falls short in providing hot water fuel savings and payback periods. This form incorporates all and since it is being offered in an excel format, can be modified after receipt for customization. This is the only calculator that incorporates all factors and savings available.

Using the Payback Period Calculation Worksheet
There are three tabs to the excel spreadsheet calculator. The first tab is the payback calculator front page where the most data entry occurs. The second tab is designated for installation estimates and the third is embedded with payback calculator formulas. Care should be taken when using the payback calculator as the form is given to you unprotected so that you can customize as you wish after receipt.

It is highly recommended that a back-up copy be made before performing any enhancements or changes. All of the yellow highlighted cells are data entry cells and all of the blue highlighted cells are formula based answer or result cells. The green highlighted cells are payback duration results. Throughout the calculator, you will find red letters next to cells so that you can track where the numbers go should they be used again elsewhere. If you are putting together a system analysis that requires an estimate taking more room than second tab spreadsheet allows, an entire estimate can be performed separately and inserted in the total cost cell as necessary.

The payback calculator has been designed to calculate the payback of any fixture, equipment or system but it needs to be noted that when using the calculated savings, it is per unit. In other words, if you are computing the payback period of replacing 3.5 GPF water closets with 1.28 GPF, the savings calculated are for one fixture replacement only. If the total amount of projected water savings is desired, the savings calculated must be multiplied by the total number of fixtures being replaced. The payback period is the same regardless of the quantity of similar fixtures being replaced.

Utility Rates
To begin the process, enter the current utility rates at the top of the form for water, sewer, electrical and gas in the yellow highlighted cells (A-D) as well as the submission title. This data entry is obtained from the owner and reflects their current rate being paid for these services over the last year. If any of the utility rates have differing summer and winter rates, these rates need to be averaged or pro-rated as necessary to reflect an accurate payback calculation that will be in use over the entire year.

Water Usage
The next step is inserting the current water usage of the existing fixture or equipment and the same for the proposed (new) fixture and equipment in Section 1 and 2. The spreadsheet will then calculate your water usage for existing (E) and new (F) and calculates the water savings in Section 3 (G). This is where the importance of knowing the building occupancy as it is necessary in determining the

fixture use rates and as a result the projected savings. Another important factor here is the number of days the building is occupied (H). If the building is a 24 hours a day, 7 days a week building, this plays a huge role in reducing the payback period duration. Once the savings per day in gallons is calculated, the spreadsheet then converts the savings to CCF (One Hundred Cubic Feet—the first 'C' is the Roman numeral 'C' standing for one hundred), which is the increment in which the utility companies bill their customers (I). Using the utility rates entered on (A-D), the spreadsheet then calculates to dollar savings for water and sewer (J) and (K) respectively.

Operation Days Matter
When determining the payback period, the number of operation days that a facility is in use makes a huge difference. Payback period is greatly influenced by the number of times the equipment is used. Consider the difference between an office building and a hospital. An office building is usually operational only 260 days a year, closed on nights and weekends. A hospital is open 365 days a year, 24 hours a day. The payback period for hospital fixtures will be significantly lower with an extra 105 days a year and extended hours.

The next two sections of the spreadsheet, 7 and 8, are only to be used in calculations where hot water is being saved. Section 7 should be utilized when the building's hot water is being generated by a natural gas or electric hot water heater. Section 8 should be utilized when the building's hot water is being generated through a steam converter and the steam is created by either natural gas or electricity. Depending upon the building, the resulting savings are shown in (L-O) cells. Section 9 is used when natural gas or electricity (used for any means) is being saved in the process unrelated to hot water but included with the efficiency or system upgrade being performed. The resulting savings are shown in (P) and (Q).

Maintenance

Section 10 is reserved for anticipated annual maintenance savings. The benefits of providing a water efficient upgrade is not only in the water, sewer and hot water fuel savings, but also maintenance. New products perform and work better, thus saving you maintenance fees. A perfect example of this is the costs of an older dishwasher requiring regular repair work to remain operational. Brand new equipment will not require anything other than basic operational maintenance and the savings difference should be reflected in this section. The savings could also include the cost avoidance of the shut-down time absorbed during the repairs if profit-generating weekend rentals had to be cancelled or emergency back-up equipment rented.

Utility Forecasting

The water, sewer, natural gas, and electrical utility rate increases anticipated are reflected in Section 11. Typically, the annual increase for each utility is independent of each other so the form has allowed for these anticipated increases to be entered separately. While no one can be expected to predict the future of these rates, it's safe to assume the rate of increase will not be zero. In fact, depending upon how much historical utility information you collect from your client, the more accurate these predictions will be. For instance, if the historical utility data shows the sewer utility rates increasing an average of 10% over the past 5 years, there is no reason not to anticipate this trend to continue after the efficiency upgrades. After applying this same reasoning to all of the utilities the form then determines the weighted average for the payback duration calculations.

Rebates

Since not all projects have rebates available and even projects with rebates available may not have them for all of the efficiency upgrades proposed, the payback duration without rebates are calculated in Section 12. For more detailed information, the payback durations are broken down individually (water, sewer, natural gas, maintenance) and then cumulatively. The total payback period calculated is the number shared with the owner in the Water Audit Report.

Sections 13 through 16 are for entering any anticipated rebates. Again, these are separated by utility as they may be awarded from anyone of them depending upon the work, your locale, and unique circumstances, you may receive more than one for the same project. An example of this would be an equipment upgrade that not only saves water and sewer, but also electricity in generating the hot water and as a result, the water utility offers a rebate for the water savings while the energy utility offers a rebate for the electrical savings.

Following the rebates data entry, Section 17 simply recalculates the installation cost incorporating them which results in a reduced final cost.

Final Payback Calculations

In Section 18, the payback durations are calculated again except for this time the payback period incorporates the revised cost from rebates. This total payback period is also included in the Water Audit report and reflects the most attractive number possible for the owner's consideration.

Below is an entire look at the payback calculator spreadsheet (embed calculator at this point of the PDF).

Here are a few examples for some common upgrade installations:

1. Hospital room, replacing 3.5 GPF Water Closet with 1.6/1.1 GPF Water Closet
2. Office building, replacing 1.0 GPF Urinal with a .0125 GPF Urinal
3. Elementary School, adding 0.5 GPM aerator to existing 2.5 GPM lavatory faucet

How to perform these examples will be discussed in detail in Chapter 5 – Water Audit Payback Period Exercises.

CHAPTER FOUR – WATER AUDIT CASE STUDIES

"Thousands have lived without love, not one without water"
- W.H. Auden

Case Study #1 Providence St. Peter Hospital in Olympia, Washington

Through its retrofit and replacement projects, Providence St. Peter Hospital in Olympia, Washington, has demonstrated that hospitals can achieve significant water savings. Hospitals consume large amounts of water, as they require water-intensive medical process equipment in addition to the typical sanitary and mechanical water use demands. However, careful planning and implementation of water-efficient practices and equipment can be quite cost-effective.

In Olympia, water rates increased 40 percent between 2000 and 2011. Realizing the potential for saving both water and operating costs, Providence St. Peter Hospital began identifying water-efficiency improvement projects as early as 1999.

Providence St. Peter Hospital is a 340-bed, 750,000-square-foot acute patient care facility. The hospital logs approximately 95,000 patient days per year and employs 2,300 staff. Ten years after the hospital began its water-efficiency efforts, staff estimated that it had reduced water use 59 percent compared to its 1998 use, or more than 31 million gallons of water saved in total. The hospital experienced water savings even though the campus grew by 17 percent and patient days increased by 22 percent between 2004 and 2009. With the help of rebates from local water and wastewater utilities, Providence St. Peter was able to afford these improvement projects and lower the facility's operating costs, all while maintaining staff and patient satisfaction.

Project Summary

As a hospital, Providence St. Peter had the opportunity to not only reduce water use in restrooms and kitchens, but also reduce water use from medical equipment. In 2001, the hospital partnered with its mechanical contractor to perform a facility water assessment and identify sources of water waste. The initial effort focused on improving operations and maintenance. Leaks can account for a large percentage of water waste in institutional facilities. As such, Providence St. Peter maintenance staff looked for potential leaks first. They analyzed irrigation systems, cooling and heating systems, faucets, kitchen equipment, hydrotherapy pool operations and other potential sources of leaks and made repairs where necessary.

Following the initial water assessment, leak detection and repair phase, Providence St. Peter began the major work of retrofitting and replacing equipment in restrooms, kitchens, and mechanical spaces, as well as medical equipment.

Most hospitals, including Providence St. Peter, use some type of sterilizer equipment to disinfect and sterilize surgical instruments, medical waste, and other materials. The hospital's four existing instrument steam sterilizers all utilized orifice venturi equipment, which uses water to produce a vacuum. By replacing the venturi equipment with electric vacuum pumps, Providence St. Peter was able to eliminate the water used to create a vacuum in these units. Additionally, piping was modified such that steam was recovered from sterilizer jacket, which would otherwise go down the drain. Steam condensate is now redirected to the boiler plant for reuse. The installation of electric vacuum pumps within the sterilizers, coupled with the steam trap maintenance program, has yielded water savings of 4,300 gallons per day (gpd).

Central vacuum pumps can also be a major source of water waste in medical facilities. These vacuum pumps can either be "dry" or "wet" based on how the vacuum seal is generated within the pump. Wet pumps use a closed impeller that is sealed with water or lubricants such as oil to generate the vacuum. Dry pumps do not use water to generate the seal for the vacuum, and therefore do not connect to a water supply. Providence St. Peter currently uses the most common type of wet vacuum pump—liquid-ring pumps, which use water to create the vacuum seal and cool the equipment—but is considering switching to dry vacuum pumps such as claw technology.

Providence St. Peter has replaced liquid-ring non-medical (control) air compressors and waste anesthesia gas pumps with non-water using equipment, resulting in savings of 2,200 and 1,400 gallons per day, respectively.

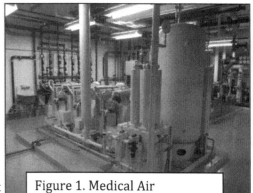

Figure 1. Medical Air

In addition to the medical equipment that was upgraded, Providence St. Peter has upgraded its sanitary fixtures. The hospital received rebates for installing dual-flush valves on existing flushometer-valve-type toilets and several new high-efficiency toilets. Providence St. Peter installed water-saving showerheads that still met patient expectations of performance. The hospital also worked with a manufacturer to design dual-flush bed pan washers—a product typically only used in the medical sector.

In the kitchen, Providence St. Peter removed the garbage disposal and instead utilizes a food separator and compost food scraps. This single change in operations has saved over 1,000,000 gallons of water per year that would have been used to flush food waste down the drain.

The facility also focused on eliminating single-pass cooling — specifically in air conditioning units and ice machines. This project alone saved 900 gallons per day, or more than 300,000 gallons of water per year.

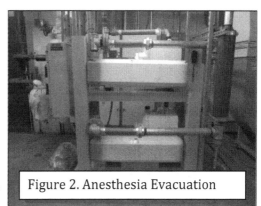

Figure 2. Anesthesia Evacuation

Providence St. Peter has additional water-efficiency projects that it is considering, including: collecting and reusing rainwater; installing submeters to better monitor water use; reducing and/or re-using clean dialysis reject water; and collecting and using air handler condensate as cooling tower make-up water.

Cost and Savings Summary

From upgrading the existing steam sterilizers, Providence St. Peter Hospital was able to reduce facility water use by 4,300 gpd or approximately 1,600,000 gallons per year. To finance the $30,232 retrofit project, the hospital received a 50 percent grant from its wastewater utility. As a result of this grant, the payback period for the project was less than 2 years.

These projects have lowered operating costs and reduced the burden on the hospital's budget; Providence St. Peter Hospital estimates that it saves approximately $140,000 each year in water and sewer bills. The hospital realized cumulative savings of $1.5 million and

31 million gallons of water between 1999 and 2011.

In addition to the cost savings benefits, patients and staff have expressed their satisfaction with the upgrades and hospital staff has noticed an increase in productivity due to better performance of the upgraded medical equipment.

Table X provides a summary of the water savings from all of the projects implemented at Providence St. Peter through 2010.

Table X. Retrofit and Replacement Project Water Savings

Project	Water Savings (gallons per year)	Payback (years)
Mechanical System Retrofits and Replacements		
Single-pass cooling equipment replacement (including air-cooled ice machines and air conditioning and refrigeration equipment)	330,000	1.1
Sanitary Retrofits and Replacements		
Flushometer-valve-type toilet retrofit (installing dual-flush valves and handles)	2,300	4.8
Urinal replacement (installing pint flush urinals)	9,900	3.4
Flushometer-valve-type toilet replacement (replacing some toilets with dual-flush models)	9,900	6.8
Showerhead replacement (installing 1.5 gallons per minute showerheads)	3,700	2.1
Faucet retrofit (reducing faucet flow rate)	1,500	4.5
Commercial Kitchen Retrofits and Replacements		
Dishwasher replacement (replacing tunnel washer with efficient model)	660,000	18
Medical Equipment Retrofits and Replacements		
Steam sterilizer pump replacement and condensate collection	1,600,000	1.9
Medical air compressors (replacing liquid-ring with a non-water using reciprocating system)	800,000	5.0
Waste anesthesia gas pump replacement (replacing liquid-ring equipment a with non-water using type)	510,000	5.7
Total	**4,200,000**	

Water Audit Report
November 19, 2009

Garfield Elementary School
325 Plymouth St NW
Olympia, WA 98502-4986

INTRODUCTION

Project Objectives / Investigation Scope

Troy Aichele reviewed Centennial Elementary School and Garfield Elementary School with Sarah Davis, Olympia School District Resource Conservation Manager on Friday, September 18, 2009 to perform a building water audit for each school. This effort was conducted in an effort to review water-saving rebate opportunities eligible for the LOTT (Lacey, Olympia, Tumwater, and Thurston County) WaterSmart Water Conservation Rebate program.

DESCRIPTION OF EXISTING FACILITIES

Fixture	Quantity
Manually Operated Lavatory Faucets (2.5 gpm)	10
Manually Operated Lavatory Faucets (0.5 gpm)	24
Floor Mounted Tank Type Water Closet (3.5 gpf)	3
Wall Hung Water Closet (3.5 gpf)	27
Wall Hung Urinals (1.0 gpf)	13
Shower (2.5 gpm)	1
Classroom / Work Area Sink (3.0 gpm)	7
Classroom / Work Area Sink (2.5 gpm)	1
Janitor Utility Sink	4
Kitchen Sink – Pot Fillers (2.5 gpm)	2
Kitchen Spray Nozzle	1
Dishwasher	1
Freezer	1

Landscaping

Garfield Elementary School – Currently no utility water is being used for landscaping.

Additional Observations

Rain Barrels were being used at the Garfield School for their garden, which in turn eliminated any need for utility water for this purpose. Due to the school being unoccupied and not maintained over the summer, the rain barrels are only useful during the school year and therefore not recommended for increased utilization.

Kitchens

The kitchen contains (2) Kitchen Sink pot fillers, Spray Nozzle, Dishwasher, and Sink. We are only recommending flow restrictors on the Sink faucets as all of the other Kitchen plumbing fixtures do not represent a reasonable payback opportunity for the cost of replacement (Dishwasher, Spray Nozzle) or does not make reasonable sense to replace (low flow pot filling faucets will simply make it longer to fill pots or the sink) and the Freezer (condensate recovery possibilities are too small in quantity to consider).

HISTORICAL USE

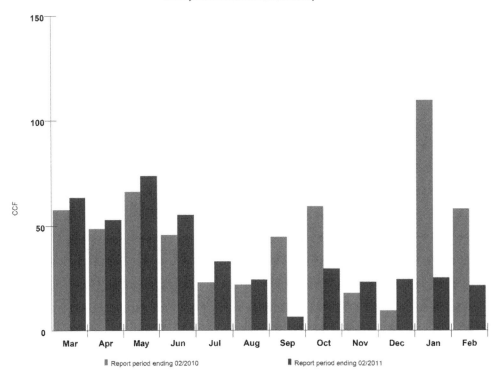

Monthly Water Use for Garfield Elementary

Centennial Elementary School (Year Ending 7/2009):
- Water: 577 CCF
- Sewer: 577 CCF
- Irrigation: 282 CCF

UTILITY RATES

- Water (Winter): $1.46/CCF
- Water (Summer): $2.91/CCF
- Sewer: $5.41/CCF

CONSERVATION OPPORTUNITIES AND RECOMMENDATIONS

1. Replace (3) three existing 2.5 gpf (average) floor mounted tank type water closets with 1.28 gpf floor mounted tank type High Efficiency Toilets (HET's). All figures below are based upon a student body and faculty total of 379 and 44 respectively, 50/50 split between male and female occupants and 3 uses / day for females and 1 use / day for males divided by all school water closets available for use. All costs quoted are without WSST.

   - Anticipated water savings: <u>102 gallons / day (24.6 ccf / year)</u>.
   - Estimated cost without rebate at $654/each: $1,962.
     - Estimated payback period without rebate: 10.52 years.
   - Estimated cost at $311/each with 75% rebate: $654.
     - Estimated payback period with 75% rebate: 2.63 years.

2. Replace (27) twenty-seven existing 2.5 gpf (average) wall hung water closets with 1.28 gpf wall hung High Efficiency Toilets (HET's). All figures below are based upon a student body and faculty total of 488 and 33 respectively, 50/50 split between male and female occupants and 3 uses / day for females and 1 use / day for males divided by all school water closets available for use. All costs quoted are without WSST.

   - Anticipated water savings: <u>918 gallons / day (221.4 ccf / year)</u>.
   - Estimated cost without rebate at $751/each: $20,277.
     - Estimated payback period without rebate: 12.09 years.
   - Estimated cost at $751/each with 75% rebate: $5,069.
     - Estimated payback period with 75% rebate: 3.02 years.

3. Install (8) eight 0.5 gpm flow restrictors on existing 3.0 gpm (average) classroom / work area sinks. All figures below are based upon a student body and faculty total of 488 and 33 respectively using one of these sinks once a day for 10 seconds. All costs quoted are without WSST.

   - Anticipated water savings: <u>106 gallons / day (25.6 ccf / year)</u>.
   - Estimated cost without rebate at $60/each: $480.
     - Estimated payback period without rebate: 2.48 years.
   - Estimated cost at $60/each with 75% rebate: $120.
     - Estimated payback period with 75% rebate: 0.62 years.

<u>Lunch 'n Learns</u> –

1. Conduct Water Conservation educational Lunch 'n Learns for students and faculty.

   - Once work has been completed, we would be happy to conduct a Lunch 'n Learn presentation to the students and faculty on what was done, why is was done, how much will be saved, and how each person can help make this effort even more successful by reporting leaks, turning off running faucets, turning faucets on only half-way when washing hands, etc.

WaterSmart Technology Program

Any cost effective water project will qualify for a financial incentive of up to 75% of the installed project cost. All of the above conservation recommendations are cost effective and we have included calculations for each along with their respective estimated installation costs and submittal cut sheets.

Also attached please find a blank City of Olympia / LOTT Alliance Water Smart Technology Program rebate application. Should you wish to proceed with any of this work, please advise and we will complete and ready the final submission form for your authorization signature.

As it states on the application, the City and LOTT Alliance will review the proposed project based on the information submitted and will contact you in writing regarding your eligibility and, if applicable, provide a preliminary rebate estimate. After reviewing this information, the Olympia School District would then need to decide whether to proceed with none, some or all of this work. Should LOTT Alliance approve any part of it, Olympia School District would then have 1 year as of the date of the application to complete the work.

Performance Contracting Offer
For the remaining 25% of the construction costs, we are pleased to offer the Olympia School District a performance contracting option where we would complete the work in its entirety, bill only for the amount of the rebate from the LOTT Alliance. Payment for the remaining unbilled amount of work would be invoiced and paid under a mutually acceptable re-payment plan.

Olympia High School Water Audit Report
1302 North Street SE
Olympia, WA 98501-3612

INTRODUCTION

Project Objectives / Investigation Scope

Troy Aichele reviewed Olympia High School with Sarah Davis, Olympia School District Resource Conservation Manager on Friday, October 22, 2009 to perform a building water audit. This effort was conducted in an effort to review water-saving opportunities for LOTT (Lacey, Olympia, Tumwater, and Thurston County) WaterSmart Water Conservation consideration.

DESCRIPTION OF EXISTING FACILITIES

Number of Students: 1,850
Number of Staff: 124

Plumbing and Kitchen Equipment

	Quantity
Plumbing	
Manually Operated Lavatory Faucets (2.0 GPM)	105
Manually Operated Lavatory Faucets (0.5 GPM)	35
Push-Activated Lavatory Faucets (0.5 GPM)	23
Floor Mounted Tank Type Water Closet (1.6 GPF)	2
Wall Hung Water Closet (1.6 GPF)	103
Wall Hung Urinal (1.0 GPF)	38
Locker Room Showers (2.5 GPM)	50
Individual Showers – Not used (2.5 GPM)	2
Janitor Sink	2
Science Lab Gooseneck Sink Faucets (2 GPM)	16
Home Education GE Dishwashers	3
Science Lab Americana Dishwasher	1
Kitchen	**Quantity**
Kitchen Spray Nozzle	1
Hobart Industrial Dishwasher	1
Air-Cooled Ice Makers	3
Residential Clothes Washers	2
Kitchen Sink Faucets– Pot Fillers (2.5 GPM)	9

Considerations:

We will not be recommending replacement or enhancement for the following items observed:

- Manually, and Push-Activated Lavatory Faucets: Already optimized at 0.5 GPM.
- Individual Showers: Not applicable as these showers are not used.
- Janitors Sink: Not applicable as these faucets are used for cleaning bucket operations
- Home Education GE and Science Lab Americana Dishwashers: There is currently not an existing rebate program available for dishwashers through Olympia and the cost of removal and replacement would exceed any reasonable payback scenarios.
- Kitchen Spray Nozzle: Existing Kitchen Spray Nozzle is operating and shutting-off properly. As we have yet to find an adequate low-flow Kitchen Spray Nozzle that continues to function as intended (with enough force to pre-rinse trays and plates properly) while saving water, we currently do not recommend changing this out. Should the Owner still desire to change this out, we would be happy to install one for beta-test purposes without expense.
- Air-Cooled Ice Makers: As these are already air-cooled, no water saving opportunities exist.
- Kitchen Sink Faucets – Pot Fillers: We do not recommend replacing as these faucets are used for pot filling operations and become a nuisance if changed to low-flow devices.

HVAC Systems and Equipment

The HVAC systems and equipment were not reviewed for water saving opportunities.

Landscaping

The Landscaping systems were not reviewed for water saving opportunities.

Site Specific Opportunities

Rainwater Harvesting

Rainwater harvesting could be utilized for fund-raising car washes, however since the car washes generally are performed for team sports and not to finance the school's operating budget we do not recommend adding this system for this purpose.

Lunch 'n Learns

Once work has been completed, we would be happy to conduct a Lunch 'n Learn presentation to the students and faculty on what was done, why is was done, how much will be saved, and how each person can help make this effort even more successful by reporting leaks, turning off running faucets, turning faucets on only half-way when washing hands, etc.

HISTORICAL USE

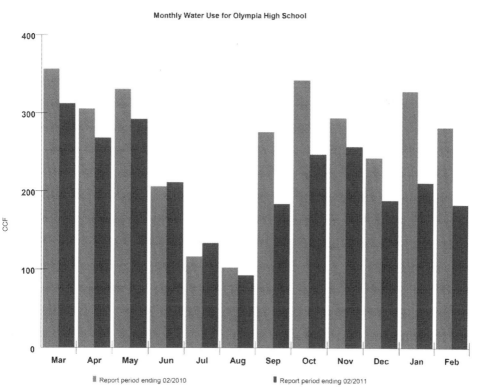

Monthly Water Use for Olympia High School

- 2008
  - Water: 3,749 CCF
  - Sewer: 3,749 CCF

- Last 5 Years (2004-2009):
  - Water Cost has increased from $1.63 to $2.22/CCF; 36% increase.
  - Sewer Cost has increased from $4.45 to $5.42/CCF; 22% increase.
  - Water use has increased from 3,337 to 3,749 CCF; 12% increase

2009 OHS UTILITY RATES

- Water (Winter): $1.60/CCF
- Water (Summer): $3.30/CCF
- Sewer: $5.42/CCF

CONSERVATION OPPORTUNITIES AND RECOMMENDATIONS

1. Install (121) one hundred and twenty-one 0.5 GPM Flow Restrictors on existing 2.0 GPM lavatories and science laboratory sinks. All figures below are based upon a student body and faculty total of 1850 and 124 respectively using one of these sinks once a day for 10 seconds. All costs quoted are without WSST.

   - Anticipated water savings: <u>290 gallons / day (72.6 ccf / year)</u>.
   - Estimated cost without rebate at $60/each: $7,260.
     - Estimated payback period without rebate: 13.04 years.
   - Estimated cost at $60/each with 75% rebate: $1,850.
     - Estimated payback period with 75% rebate: 3.26 years.

2. Replace (2) two 1.6 GPF Floor Mounted Tank Type Water Closets with 1.28 GPF Floor Mounted Tank Type Dual-Flush Water Closets. All costs quoted are without WSST.
   - Anticipated water savings: <u>17.2 gallons / day (4.2 ccf / year)</u>.
   - Estimated cost without rebate at $654/each: $1,308.
     - Estimated payback period without rebate: 39.97 years.
   - Estimated cost at $654/each with 75% rebate: $327.
     - Estimated payback period with 75% rebate: 9.99 years.

   2A. Add (2) two Flush-Choice Dual Flush Tank Type Retrofit Kits to existing 1.6 Tank Type Water Closets. All costs quoted are without WSST.
   - Anticipated water savings: <u>16 gallons / day (3.8 ccf / year)</u>.
   - Estimated cost without rebate at $146/each: $292.
     - Estimated payback period without rebate: 9.64 years.
   - Estimated cost at $146/each with 75% rebate: $73.
     - Estimated payback period with 75% rebate: 2.41 years.

3. Add (103) one hundred and three Manual Dual Flush Flushometer Retrofit Kits to existing 1.6 GPF Wall Hung Water Closets. All costs quoted are without WSST.
   - Anticipated water savings: <u>1,957 gallons / day (473.8 ccf / year)</u>.
   - Estimated cost without rebate at $85/each: $8,755.
     - Estimated payback period without rebate: 2.36 years.
   - Estimated cost at $85/each with 75% rebate: $2,188.

- Estimated payback period with 75% rebate: 0.59 years.

3A. (Alternate to Item #3 above) Add (103) One Hundred and Three Sensor Operated Solar Battery Powered Dual Flush Flushometer Retrofit Kits to existing 1.6 GPF Wall Hung Water Closets. All costs quoted are without WSST.
- Anticipated water savings: 1,957 gallons / day (473.8 ccf / year).
- Estimated cost without rebate at $357/each: $36,771.
  - Estimated payback period without rebate: 9.92 years.
- Estimated cost at $357/each with 75% rebate: $9,192.
  - Estimated payback period with 75% rebate: 2.48 years.

4. Replace (38) thirty-eight manual flush 1.0 GPF Urinals with sensor operated solar battery powered 0.125 GPF Urinals. All costs quoted are without WSST.
- Anticipated water savings: 1,972 gallons / day (414.2 ccf / year).
- Estimated cost without rebate at $1,239/each: $47,082.
  - Estimated payback period without rebate: 14.41 years.
- Estimated cost at $1,239/each with 75% rebate: $11,770.
  - Estimated payback period with 75% rebate: 3.60 years.

5. Replace (50) fifty 2.5 GPM Locker Room Shower Head internal gaskets with 1.5 GPM internal gaskets. All costs quoted are without WSST.

- Anticipated water savings: 20 gallons / day (5 ccf / year)
- Estimated cost without rebate at $11/each: $550.
  - Estimated payback period without rebate: 14.52 years.
- Estimated cost at $11/each with 75% rebate: $138.
  - Estimated payback period with 75% rebate: 3.63 years.

6. Replace (1) one existing Hobart Kitchen Dishwasher with a new Champion Model #76 DRPW Dishwasher. Costs quoted are without WSST.

- Anticipated water savings: 1,947 gallons / day (533.7 ccf / year).
- Estimated cost without rebate: $50,888.
  - Estimated payback period without rebate: 12.11 years.
- Estimated cost at $311/each with 75% rebate: $12,722.
  - Estimated payback period with 75% rebate: 3.03 years.

7. Replace (2) two Kenmore Clothes washers with a GE Energy Star / CEE Qualified Tier 1 Clothes Washer (installed by Owner). All costs quoted are without WSST.
- Anticipated water savings: 44 gallons / day (10.6 ccf / year).
- Estimated cost without rebate at $546.41/each: $1,093.
  - Estimated payback period without rebate: 13.11 years.
- Estimated cost at $546.41/each with 75% rebate: $273
  - Estimated payback period with 75% rebate: 3.28 years.

For items #3 and #3A, we strongly recommend beta-testing both models prior to selecting desired dual flush technology for the water closets.

WaterSmart Technology Program

Any cost effective water project will qualify for a financial incentive of up to 75% of the installed project cost. All of the above conservation recommendations are cost effective and we have included calculations for each along with their respective estimated installation costs and submittal cut sheets.

Also attached please find a blank City of Olympia / LOTT Alliance Water Smart Technology Program rebate application. Should you wish to proceed with any of this work, please advise and we will complete and ready the final submission form for your authorization signature.

As it states on the application, the City and LOTT Alliance will review the proposed project based on the information submitted and will contact you in writing regarding your eligibility and, if applicable, provide a preliminary rebate estimate. After reviewing this information, the Olympia School District would then need to decide whether to proceed with none, some or all of this work. Should LOTT Alliance approve any part of it, Olympia School District would then have 1 year as of the date of the application to complete the work.

Performance Contracting Offer

For the remaining 25% of the construction costs, we are pleased to offer the Olympia School District a performance contracting option where we would complete the work in its entirety, bill only for the amount of the rebate from the LOTT Alliance. Payment for the remaining unbilled amount of work would be invoiced and paid under a mutually acceptable re-payment plan negotiated between the Olympia School District and our company.

Orla Learning Academy Water Audit Report
2601 26th Avenue
Olympia, WA

April 9, 2010
REVISION 1: June 7, 2010

INTRODUCTION

Project Objectives / Investigation Scope

Troy Aichele reviewed Capital High School with Sarah Davis, Olympia School District Resource Conservation Manager to perform a building water audit. This effort was conducted in an effort to review water-saving opportunities for LOTT (Lacey, Olympia, Tumwater, and Thurston County) WaterSmart Water Conservation rebate consideration.

REVISION #1: Added Option #2A for Owner Consideration.

DESCRIPTION OF EXISTING FACILITIES

Number of Students: 389
Number of Staff: 20

Plumbing and Kitchen Fixtures

	Quantity
Plumbing	
Manually Operated Lavatory Faucets (2.0 GPM)	46
Manually Operated Lavatory Faucets (0.5 GPM)	6
Manually Operated Lavatory Faucets (2.5 GPM)	4
Floor Mounted Water Closet (3.5 GPF)	18
Wall Hung Water Closet (3.5 GPF)	4
Antique Faucets (2.5 GPM)	15
Antique Faucets (5.0 GPM)	9
Kitchen Spray Nozzles	2
Kitchen Sink Faucets – Pot Fillers	3
Janitor Service Sink	1
Shower (2.5 GPM)	1
Custom Urinals	2
Integral Wall Urinal (1.0 GPF)	2

Considerations:

We will not be recommending replacement or enhancement for the following items observed:

- Antique Faucets (2.0, 2.5 and 5.0 GPM): Several faucets at Orla would require complete replacement and/or fixture replacement in order to obtain water and sewer savings. The water and sewer savings resulting from this replacement work would not justify the cost of replacement and therefore not recommended.
- Lavatory faucets already optimized at 0.5 GPM (push and manually activated).
- Individual Shower: Not applicable as shower audited were not being used.
- Janitors Sink: Not applicable as this faucets are used for cleaning bucket operations
- Custom Urinals: These urinals would require a complet replacement and plumbing remodel and the water and sewer savings resulting from this replacement work would not justify the cost of replacement and therefore not recommended
- Kitchen Spray Nozzle: Existing Kitchen Spray Nozzle is operating and shutting-off properly. As we have yet to find an adequate low-flow Kitchen Spray Nozzle that continues to function as intended (with enough force to pre-rinse trays and plates properly) while saving water, we currently do not recommend changing this out. Should the Owner still desire to change this out, we would be happy to install one for beta-test purposes without expense.
- Kitchen Sink Faucets – Pot Fillers: We do not recommend replacing as these faucets are used for pot filling operations and become a nuisance if changed to low-flow devices.

HVAC Systems and Equipment

The HVAC systems and equipment were not reviewed for water saving opportunities.

Landscaping

The Landscaping systems were not reviewed for water saving opportunities.

Site Specific Opportunities

Lunch 'n Learns
Once work has been completed, we would be happy to conduct a Lunch 'n Learn presentation to the students and faculty on what was done, why is was done, how much will be saved, and how each person can help make this effort even more

successful by reporting leaks, turning off running faucets, turning faucets on only half-way when washing hands, etc.

WATER AND SEWER HISTORICAL USE

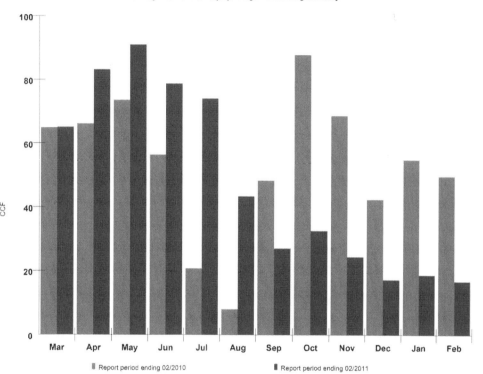

Monthly Water Use for Olympia Regional Learning Academy

- 2005: 835 CCF
- 2006: 1638 CCF
- 2007: 843 CCF
- 2008: 576 CCF
- 2009: 658 CCF

2009 CHS UTILITY RATES

- Water (Audited 2009 average): $2.09/CCF
- Sewer: $5.42/CCF

CONSERVATION OPPORTUNITIES AND RECOMMENDATIONS

1. Replace (18) eighteen Floor Mounted 3.5 GPF Water Closets with Floor Mounted 1.28 GPF Water Closets. All costs quoted are without WSST.
    - Anticipated water savings: <u>44.4 gallons / day (192.6 ccf / year)</u>.
    - Estimated cost without rebate at $431/each: $7,758.
        o Estimated payback period without rebate: 5.13 years.
    - Estimated cost at $431/each with 75% rebate: $1,939.
        o Estimated payback period with 75% rebate: 1.28 years.

2. Replace (4) four Wall Hung 3.5 GPF Water Closets with Wall Hung 1.28 GPF Water Closets. All costs quoted are without WSST.
    - Anticipated water savings: <u>178 gallons / day (42 ccf / year)</u>.
    - Estimated cost without rebate at $557/each: $2,228.
        o Estimated payback period without rebate: 6.62 years.
    - Estimated cost at $557/each with 75% rebate: $557.
        o Estimated payback period with 75% rebate: 1.65 years.

    2A. Replace (4) four Wall Hung 3.5 GPF Water Closets with Wall Hung 1.6 GPF Water Closets with 1.6/1.1 Dual Flushometers. All costs quoted are without WSST.
    - Anticipated water savings: <u>186 gallons / day (44.8 ccf / year)</u>.
    - Estimated cost without rebate at $477/each: $1,908.
        o Estimated payback period without rebate: 5.41 years.
    - Estimated cost at $477/each with 75% rebate: $477.
        o Estimated payback period with 75% rebate: 1.35 years.

WaterSmart Technology Program

Any cost effective water project will qualify for a financial incentive of up to 75% of the installed project cost. All of the above conservation recommendations are cost effective and we have included calculations for each along with their respective estimated installation costs and submittal cut sheets.

Also attached please find a blank City of Olympia / LOTT Alliance Water Smart Technology Program rebate application. Should you wish to proceed with any of this work, please advise and we will complete and ready the final submission form for your authorization signature.

As it states on the application, the City and LOTT Alliance will review the proposed project based on the information submitted and will contact you in writing regarding your eligibility and, if applicable, provide a preliminary rebate estimate. After reviewing this information, the Olympia School District would then need to decide whether to proceed with none, some or all of this work. Should LOTT Alliance approve any part of it, Olympia School District would then have 1 year as of the date of the application to complete the work.

Performance Contracting Offer
For the remaining 25% of the construction costs, we are pleased to offer the Olympia School District a performance contracting option where we would complete the work in its entirety, bill only for the amount of the rebate from the LOTT Alliance. Payment for the remaining unbilled amount of work would be invoiced and paid under a mutually acceptable re-payment plan negotiated between the Olympia School District and our company.

CHAPTER FIVE – HOW TO WRITE A WATER AUDIT REPORT

"If there is magic on this planet, it is contained in water"
- Loran Eisely, The Immense Journey, 1957

Writing the Water Audit Report is the final step. While everyone will have their own style and company influences in writing proposal letters, the attached is a guideline to follow in creating a Water Audit Report using one of the case studies as an example. Comments in this section will be made in blue.

Title: The name, address of the facility where the Water Audit was performed.

NAME OF FACILITY Water Audit Report
Address Line 1
Address Line 2

Month, Day, Year

Introduction: The introduction is the opening of your Water Audit Report and simply an overview of the audit objectives and who performed the actual water audit.

INTRODUCTION

Project Objectives / Investigation Scope

YOUR NAME from YOUR COMPANY NAME reviewed NAME OF FACILITY on MONTH, DAY, YEAR in an effort to review water, waste-water and hot water conservation opportunities for NAME OF OWNER.

Description of existing facilities: It is advised to keep this short and simple as the person you are sending this to already knows the size of their facility and what it looks like. They are looking for water conservation opportunities and your Water Audit report should remain focused on just that. It is, however, important to reiterate the number of staff, student and employee numbers as these are used throughout your Water Audit Report in calculating your payback periods.

DESCRIPTION OF EXISTING FACILITIES

Number of STAFF/STUDENTS/EMPLOYEES: XXX

An itemized list of all of the fixtures reviewed is essential. Even if all of the items listed will not become replaced as a result of the Water Audit, it is important to list them all to show the Water Audit was thorough and complete. It can also be used as a future guide for follow-up Water Audits should they ever become desired. And perhaps most importantly, this also gives the Owner a useful snapshot view of the total number of items and fixtures throughout their facility.

Plumbing and Kitchen Equipment

	Quantity
Plumbing	
Manually Operated Lavatory Faucets (2.0 GPM)	121
Manually Operated Lavatory Faucets (0.5 GPM)	35
Push-Activated Lavatory Faucets (0.5 GPM)	23
Floor Mounted Tank Type Water Closet (1.6 GPF)	2
Wall Hung Water Closet (1.6 GPF)	103
Wall Hung Urinal (1.0 GPF)	38
Individual Showers – Not used (2.5 GPM)	2
Kitchen	**Quantity**
Hobart Industrial Dishwasher	1
Air-Cooled Ice Makers	3
Kitchen Sink Faucets– Pot Fillers (2.5 GPM)	9

It is just as important to list the items not being suggested for improvement as is listing the ones being suggested. After listing all of the recommended items, summarize the ones being bypassed and explain why. The owner will appreciate your due diligence along with looking out for their financial interests in not recommending anything that doesn't make sense payback period-wise or would result in the fixture or system that is less practical or user friendly. The following are a few examples of this.

Considerations:
We will not be recommending replacement or enhancement for the following items observed:
- Manually, and Push-Activated Lavatory Faucets: Already optimized at 0.5 GPM.
- Individual Showers: Not applicable as these showers are not used.
- Air-Cooled Ice Makers: As these are already air-cooled, no water saving opportunities exist.

- Kitchen Sink Faucets – Pot Fillers: We do not recommend replacing as these faucets are used for pot filling operations and become a nuisance if changed to low-flow devices.

While fixture replacements will typically always be the low hanging fruit for water conservation, there remain numerous opportunities in improving HVAC systems and equipment. Should you have the opportunity to perform a system comparison or replace an aging piece of major HVAC equipment (for example, cooling towers) this is the place to write about what you observed and what you're proposing.

HVAC Systems and Equipment

The HVAC systems and equipment were not reviewed for water saving opportunities.

Landscaping may be your client's largest water consuming system in their facility. If you have found ways to improve this system or would like to suggest completely removing this system while recommending planting indigenous plants, the following paragraph is where you would want to make these suggestions.

Landscaping

The Landscaping systems were not reviewed for water saving opportunities.

Every building is different and often times you will be faced with opportunities that fall outside of most typical parameters and checklists. Perhaps rainwater harvesting will be a great fit for this facility. Or maybe the local utility is expanding their reclaimed water system to your area with new user rebates. The list here could be endless and it's simply up to you during your Water Audit review to keep your eyes and ears open to how the facility uses its water and existing opportunities available in its utility district.

Site Specific Opportunities

Rainwater Harvesting
Rainwater harvesting could be utilized for car washes, however our calculated return on investment for this work was over 10 years and as a result we are currently not recommending the addition of this system.

In an effort to share with the owner their historical use and future increase trends, including this in your Water Audit Report is helpful. Some owners keep great records of their utility usage and others may not and still others who may be involved in the financial decision in funding the project may be unaware of the current usage, rates and past cost trends. Spelling it out on the Water Audit Report ensures everyone is on the same page in this regard and helps validate utility cost increases anticipated in your payback calculations performed.

HISTORICAL USE

- YEAR
    - Water: XXX CCF
    - Sewer: XXX CCF

- Last X Years (XXXX-XXXX):
    - Water Cost has increased from $X.XX to $X.XX/CCF; XX% increase.
    - Sewer Cost has increased from $X.XX to $X.XX/CCF; XX% increase.
    - Water use has increased from X,XXX to X,XXX CCF; XX% increase

It is important to know your rates during summer and winter. Most places have different rates during different times of the year which is important in calculating the payback period. As a prime example, schools typically use the least amount of water during the summer months for indoor use, but may use the most for irrigation during this same time period. If you are proposing an irrigation upgrade, you may be able to use the highest water rate during the summer for your payback calculations, but when figuring indoor water use, you may need to only use the winter rates or an average of the two.

2009 FACILITY NAME **UTILITY RATES**

- Water (Winter): $X.XX/CCF
- Water (Summer): $X.XX/CCF
- Sewer: $X.XX/CCF

The most important part of your water audit is next: recommendations and their associated payback periods. This 'grocery list' gives the owner a chance to pick and choose which conservation ideas they would like to proceed with or go with the entire list at once. Either way, by listing each item individually, you've provided a transparent, easy to follow guide from which to chose from for their facility.

CONSERVATION OPPORTUNITIES AND RECOMMENDATIONS

1. Install (121) one hundred and twenty-one 0.5 GPM Flow Restrictors on existing 2.0 GPM lavatories faucet sinks. All figures below are based upon a STAFF/STUDENT/EMPLOYEE total of XXX using one of these sinks XX times a day for XX seconds. All costs quoted are without WSST.

    - Anticipated water savings: XX gallons / day (XX.X CCF / year).
    - Estimated cost without rebate at $XX/each: $X,XXX.

- Estimated cost with XX% rebate: $X,XXX.
  - Estimated payback period with XX% rebate: X.XX years.

More often than not, the building owner is unaware of current utility rebates available for their facility. Even though they have probably been made aware of the programs available during your initial consultations or Water Audit walk-through, it's important to reiterate and put into writing their currently available opportunities.

Name of Rebate Program

Describe how the rebate program(s) work in this paragraph.

Example:
According to the above water conservation program, qualifying work shall be considered for incentives up to 50% of the installed cost. Should this work be desired, we would be happy to submit our calculations and submittal cut sheets of the recommended products for this consideration.

The following paragraph is simply a reminder of additional services some companies may be able to provide. Whether or not your company would like to offer this service is up to your company. If you do decide to offer this service, the following is an example of how to present this offering.

YOUR COMPANY NAME Performance Contracting Offer

For the remaining XX% of the construction costs, YOUR COMPANY NAME is pleased to offer a performance contracting option where YOUR COMPANY NAME would complete the work in its entirety and bill only for the amount of the rebate. All remaining unbilled work would be invoiced and paid under a mutually acceptable re-payment plan negotiated between the NAME OF OWNER and YOUR COMPANY NAME.

Sincerely,

YOUR COMPANY NAME

Your Name
Your Title

CHAPTER SIX – TIPS, STRATEGIES AND LESSONS LEARNED

*"My books are like water; those of great geniuses are wine.
Fortunately, everybody drinks water"*
    - Mark Twain

Every project brings unique situations and challenges. This chapter will outline tips, strategies and lessons learned from actual Water Audits and Installations performed.

I.	Getting on a facility's green team
II.	Problems with unique fixtures
III.	Post installation flushometer adjustments
IV.	Infectious control
V.	Water audit ticksheet diligence
VI.	Mixing plumbing fixture parts with differing performance characteristics
VII.	Fixture footprints
VIII.	System Freezing vs. shut down
IX.	Importance of Beta-testing
X.	Problematic existing conditions
XI.	Aerators vs. flow controls
XII.	Straying from recommendations
XIII.	Thinking outside the box
XIV.	Small Water Closet water spots
XV.	Dual flush tank type kits
XVI.	Additional use stickers
XVII.	Sacrificing systems performance over water efficiency
XVIII.	The importance of system water pressure
XIX.	Stop looking for new products
XX.	Follow up audits.

With every building being different, there will always be a new situation requiring new analysis and solutions. That being said, there are common situations that continue to arise and in an effort to avoid some mistakes already made, they have been summarized for you in the following.

I.   Getting on a facility's green team – Every facility interested in improving their building's efficiency has one of these. If you can find a way to get on their team, great, but if not, don't stop there. All you really need is to find out who is on the green team and get to know one of them so that you can introduce them to the concept of water auditing, saving potential and available rebates. Perhaps you could get a chance to talk about water auditing at one of their lunches or meetings.

II.   Problems with unique fixtures – While this can happen in building, young or old, more often than not this is experienced in older structures and the important thing to remember is that unique fixtures come with unique (and expensive) solutions that will hurt your payback period calculations. For example, unique water closets may require additional plumbing rework before (or after) changing the fixture out; older faucets may require the entire faucet to be changed out where as a new faucet only requires an aerator change.

III.   Post-installation flushometer adjustments – While most all of the new water efficient fixtures work well, they also need to be tested and sometimes adjusted to work correctly. In particular, the dual flushometers work best when adjusted after installation. Skipping this step may leave a user frustrated with having to flush more often when the device was supposed to save water.

IV.   Infectious control – Specifically for those looking to perform upgrades in medical facilities or in health care, this can be concerning for those operating the facility. It is a common misconception that less water means less cleansing, however with customer education, documentation and proper product selection, this concern can be overcome. In one of our facilities, the infection control was the final hurdle for replacing 0.125 gallon per flush urinals. They were concerned that with less water per flush, there wouldn't be enough water to cleanse the fixture after each use. To address their concerns, we filmed and photographed the urinal in mid-flush with an infrared camera which proved that the back of the urinal was being scoured by the water during every flush by virtue of the color differences apparent under infrared. Other fixtures or systems may have different infection control concerns and with patient safety being priority one in healthcare, we need to do all we can to make sure anything we replace or upgrade will not sacrifice patient care.

V.	Water audit ticksheet diligence – The ticksheet is the key to counting, documenting, purchasing and installing the upgraded fixtures in a building. It has to be right. It's also easy to lose track, get in a hurry or leave notes that may be misinterpreted later by yourself or others later. This can easily lead to incorrect payback calculations and incorrectly ordering replacement fixtures. This will cost additional time, material and money not included in your installation estimates. It is in everyone's best interest to be as diligent as possible in the creation of the ticksheet and it is helpful to understand that something that seems clear during the walkthrough may not be so later when you get back to the office. Pictures are also important to take should you encounter anything out of the ordinary.

VI.	Mixing plumbing fixture parts with differing performance characteristics – One of the most common reasons of water closets not flushing correctly during a water audit review is a facility that changed out all of their 3.5 GPF flushometers with 1.6 GPF flushometers but never changed out the bowls. The water closet bowls are engineered and designed to for specific flush quantities and lowering the flushometer water volume without changing the bowl typically results in constant clogging. As far as the water audit is concerned this becomes a difficult fixture to replace with an attractive payback period as they already have a 1.6 GPF fixture, and the savings projection by changing the bowl will only be from fewer flushes and maintenance savings which are difficult to estimate.

VII.	Fixture footprints – When replacing wall hung or floor mounted fixtures, it's important to perform a quick double-check of the fixture footprint to confirm the replacement fixture is larger than the existing fixture. This is important because if the footprint of the new fixture is smaller than the existing, wall and floor work will then become necessary and add to the costs of construction (and raising your payback period duration).

VIII.	System Freezing vs. shut down – One of the difficulties with making some of the plumbing and piping changes is having to shut-down the system service the fixture or equipment for a set period of time in order to perform the change-out. One of the ways around this is to freeze the line in lieu of shutting it down and draining it. While this still creates a situation where the line to the fixture is not available, it saves having to shutdown any adjacent fixtures downstream of the shutoff valve. A word of caution goes with this however. Whoever is doing the line freezing needs to be properly trained in this procedure because if you lose a frozen line, you will have an open line flowing freely. This can cause untold damage depending upon when and where you are performing the work.

IX.   Importance of Beta-testing – Even with MaP testing available for most fixtures, nothing can replace actually testing your chosen product in your building. The user can test it, feel it and believe it by the time you are done with a live test. Even more important, if the product doesn't meet their needs or expectations, a replacement can be tested. Water quality and type can vary wildly from area to area and you should not assume a product that worked well in one building would work well in another. Beta-testing is also an insurance plan as you do not want to change out an entire facility with a certain product before fully knowing it works as desired and intended.

X.   Problematic existing conditions – If there are existing plumbing or piping complications or problems, they need to be corrected prior to performing and fixture or equipment upgrades. As an example, if drain carry problems exist with a particular waste line, lowering the water flow with highly efficient fixtures will only exacerbate the problem. It is in these situations where we identify the problem, and (for that reason) explain why improvements are not recommended for that particular area or fixture in the Water Audit Report.

XI.   Aerators vs. flow controls - Both of these devices cost about the same, but if you are performing health care work it is important to know (and let the owner know that you know) that aerators are not allowed in health care facilities per code. It is also important to know for installation labor reasons—some faucets are more difficult (or may not even be able to accommodate) flow controls.

XII.   Straying from recommendations – As the Water Auditor, we recommend products, and associated installation estimates, for specific pieces of equipment we know works well, offers minimal maintenance and can be installed efficiently. Should the owner use the Water Audit Report as a guide and decide to choose different products, manufacturers and model number for opportunities presented, they should know the savings, installation and maintenance projections included in your Water Audit Report are no longer valid. Despite the progress of water efficient products available in the marketplace, the fact remains they all do not work equally well and the owner should be made aware of this fact.

XIII.   Thinking outside the box – Every building is different and often times so are the rebate plans and owner funding availability. Payback period thresholds, rebate requirements and the cost of work can dictate whether a project moves forward or not. Sometimes it may make sense to have the owner purchase all of the fixtures and leave the contractor only in charge of the installation. Perhaps the rebate can help offset the purchase of the fixture purchases or fixture installations. Most rebate programs are fairly new and may have flexibility in how their rebate dollars are awarded. Talk to the utility offering the rebate and find out if there are any creative ways to maximize rebates available as a way to lower the duration of the payback period.

XIV. Small Water Closet water spots – MaP testing is a fantastic way to get started but fair warning it does not tell you everything about water closets (another reason why beta-testing is so important). Specifically, the MaP testing will not tell you the size of the water closet water spot. The water spot is the surface area of water within the toilet bowl. The smaller the water spot, the more water closet maintenance may be required which may lengthen your payback period or worse leave the owner dissatisfied with the fixture replacement.

XV. Dual flush tank type kits – There are plenty of these devices on the market today however it has proven nearly impossible to find one that is cost effective and works effectively offering an attractive payback period. Obviously, if the upgrade kit doesn't work, it shouldn't be done. If the upgrade kit costs too much or takes too long to install, you might as well upgrade the entire fixture.

XVI. Additional use stickers – Most all dual-flushometers come with their own signs on how the dual flush operates. Up for small flush, down for large flush but adding a customized sticker of choice indicating the same flushing parameters has proven effective in many of our change-out facilities. It's certainly not necessary, but for those owners interested in maximizing their savings and shortening the occupancy learning curve, additional fixture stickers are a wise choice.

XVII. Sacrificing systems performance over water efficiency – Kitchen faucets used for pot fillers are a good example of this lesson learned. Yes, you can upgrade these faucets to a lower flow, however if all they are ever used for is filling pots with water for cooking purposes, they will use the same amount of water regardless of the water flow per minute and in fact will become an irritation to the user because they will have to wait longer for the pot to fill.

XVIII. The importance of system water pressure – Plumbing fixtures and equipment all have water pressure ranges where they function best. As a Water Auditor, we need to pay particular attention to the existing water pressure prior to making any changes. If the water pressure is too high, showers may use more water than anticipated. If the water pressure is too low, the flushometers may not operate correctly.

XIX.   <u>Stop looking for new products</u> – New, better, less expensive and more versatile products are becoming available on a monthly basis. As a water auditor it is our responsibility to stay on top of this market and know what is available to our customers as not only will they save more operations and maintenance dollars, but also, if the newer equipment is easier to install or comes at a less expensive price the end result will be a shorter payback period.

XX.   <u>Follow up audits.</u> – Dovetailing from the idea of never stopping to look for new products is the fact that if after two years you have not gone back through a facility, you should because there will be so many new products available for consideration (not mentioning rising utility rates) that a product or system upgrade that didn't offer an attractive payback period two years ago, just might today. Rebates may also change as well which in turn would also drastically change the payback period and allow an upgrade project to move forward. Lastly, the owner may not have had funding two years prior and now they do.